T4-ADW-625

SPOTLIGHT ON GLOBAL ISSUES

THE ENDANGERED SPECIES PROBLEM

Daniel R. Faust

ROSEN PUBLISHING

New York

Published in 2022 by The Rosen Publishing Group, Inc.
29 East 21st Street, New York, NY 10010

Copyright © 2022 by The Rosen Publishing Group, Inc.

All rights reserved. No part of this book may be reproduced in any form without permission in writing from the publisher, except by a reviewer.

First Edition

Editor: Theresa Emminizer
Book Design: Michael Flynn

Photo Credits: Cover Gudkov Andrey/Shutterstock.com; (series globe background) photastic/Shutterstock.com; p. 4 WishnclickS/Shutterstock.com; p. 5 Sushaaa/Shutterstock.com; p. 6 Nenov Brothers Images/Shutterstock.com; p. 7 Shannon Dunaway/Shutterstock.com; p. 9 (cicada) Toru Kimura/Shutterstock.com; p. 9 (deer) marcin jucha/Shutterstock.com; p. 9 (trout) Harel Bartik/Shutterstock.com; p. 10 Ikordela/Shutterstock.com; p. 11 Fiona Ayerst/Shutterstock.com; p. 12 Lukasz Pawel/Shutterstock.com; p. 13 Popperfoto/Getty Images; p. 15 Herschel Hoffmeyer/Shutterstock.com; p. 16 Saul Loeb/AFP/Getty Images; p. 17 https://commons.wikimedia.org/wiki/Category:Phocoena_sinus#/media/File:Vaquita4_Olson_NOAA.jpg; p. 19 TigerStock's/Shutterstock.com; p. 21 (finches) UniversalImagesGroup/Getty Images; p. 21 (Darwin) GraphicaArtis/Archive Photos/Getty Images; p. 22 Rich Carey/Shutterstock.com; p. 23 John Carnemolla/Shutterstock.com; p. 25 Dmytro Zinkevych/Shutterstock.com; p. 26 Portland Press Herald/Getty Images; p. 27 RPBaiao/Shutterstock.com; p. 29 Bloomberg/Getty Images.

Library of Congress Cataloging-in-Publication Data

Names: Faust, Daniel R., author.
Title: The endangered species problem / Daniel R. Faust.
Description: New York : Rosen Publishing, 2022. | Series: Spotlight on
  global issues | Includes bibliographical references and index. |
Identifiers: LCCN 2019057402 | ISBN 9781725323247 (paperback) | ISBN
  9781725323278 (library binding) | ISBN 9781725323254 (6 pack)
Subjects: LCSH: Endangered species--Juvenile literature. | Extinction
  (Biology)--Juvenile literature. | Mass extinctions--Juvenile literature.
  | Wildlife conservation--Juvenile literature.
Classification: LCC QL83 .F38 2022 | DDC 333.95/22--dc23
LC record available at https://lccn.loc.gov/2019057402

Manufactured in the United States of America

Some of the images in this book illustrate individuals who are models. The depictions do not imply actual situations or events.

CPSIA Compliance Information: Batch #CSR22. For further information contact Rosen Publishing, New York, New York at 1-800-237-9932.

Find us on

CONTENTS

BEE CONCERNED 4
FOOD, SHELTER, AND TOOLS 6
WHAT ARE THE SPECIES IN YOUR NEIGHBORHOOD? 8
WHEN GONE IS GONE 12
MASS EXTINCTION EVENTS 14
WHAT MAKES A SPECIES ENDANGERED? 18
HUMAN IMPACT 22
SAVING ENDANGERED SPECIES 24
THE ENDANGERED SPECIES ACT 28
THEY CAN BE SAVED! 30
GLOSSARY 31
INDEX 32
PRIMARY SOURCE LIST 32
WEBSITES 32

CHAPTER ONE

BEE CONCERNED

Between 2006 and 2007, beekeepers around the world noticed something strange. Their hives were mysteriously empty. Domesticated honeybees, or honeybees bred and raised for use by people, weren't the only bees disappearing. Many species of wild bees were also disappearing. Scientists coined the term "colony collapse disorder," or CCD, to describe this phenomenon. Colony collapse disorder occurs when the worker bee population of a bee hive, or colony, disappears, while the queen and the young bees remain. A bee hive without worker bees will eventually die.

Worker bees do most of the work in a hive. They gather food, make honey, defend the hive, tend to the queen, and take care of the young. Without worker bees, a hive won't survive for very long.

Colony collapse disorder is dangerous. If a bee species disappears completely, it's gone forever. Why is this important? People use bees for more than honey. Many farmers use bees to pollinate their crops. If honeybees disappear, it might become harder for farmers to grow enough food, which could lead to famine, or food shortage. Wild bees also play an important role in pollination. These bees help wild plants reproduce. Other animals use these plants for food or shelter. If wild bees disappear, these plants might also disappear, and so would the other species that need those plants to survive.

Bees and plants rely on each other to survive. Many species rely on other species to survive, even humans. If one of those species dies out, it becomes extremely likely the other species will too. This is why it's important to protect all the different species of plants and animals on Earth.

CHAPTER TWO

FOOD, SHELTER, AND TOOLS

Humans also rely on other species to survive. In prehistoric times, humans gathered food from whatever plants and trees they could find where they were living or traveling. Eventually, humans began planting and raising these plants and trees on their own. Agriculture changed the way humans lived.

In addition to relying on plants and trees for some food, humans also rely on plants for some medicines. The aloe plant, for example, produces a substance that soothes burns on the skin. Scientists are studying plant species around the world, searching for new cures for diseases such as cancer and Alzheimer's disease. Many of the plants that scientists are studying are native to habitats that are in danger of being destroyed, including the tropical rain forest. If these plants go extinct, we'll never be able to benefit from what they have to offer.

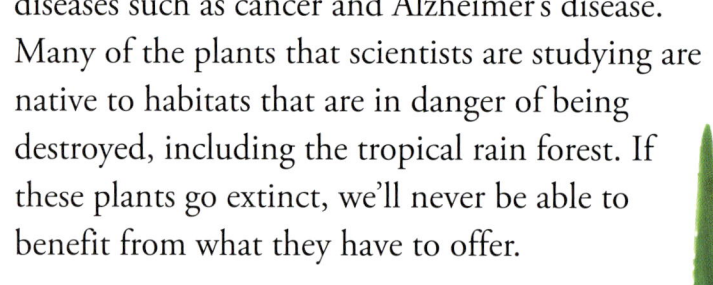

ALOE PLANT

A tepee, or tipi, is a tent made out of wood and leather. Tepees were used by the indigenous peoples of the plains of North America, notably the Sioux, Blackfeet, and Crow nations.

In addition to food and medicine, we use plants and animals for creating shelter and clothing. Humans have used timber from trees to build houses and make fires to cook food, stay warm, and keep predators away. We've used the fur from animals to make clothes and blankets to protect us from the elements. Leather, which humans use to make clothing, comes from the hide, or skin, of animals such as cattle and deer. Many cultures have also used animal hides to build shelters, such as the tents made by some Native American tribes. Cloth such as cotton or linen is made by weaving together plant fibers. Humans have also used the bones, horns, and antlers of some animals to make tools and weapons.

CHAPTER THREE
WHAT ARE THE SPECIES IN YOUR NEIGHBORHOOD?

One description of the word "community" is a group of people who share customs, values, identity, or religious beliefs. Communities often also share a given geographical area, such as a country, city, village, town, or neighborhood. Human communities can also include distinct social groups at a school, workplace, or religious institution. For example, you and your neighbor might go to different schools. While you both belong to one community, the community of people who live on your street, you also belong to different school communities. Although we often talk about communities in terms of human social groups, there are also ecological communities.

An ecological community is a group of all the species living in the same place. If you think of your neighborhood as a community that includes all the people and families that live in the same place as you, then it's easy to think of an ecological community as a natural neighborhood where different species of plants, animals, and other organisms live. An ecological community relies on a balance of nature, meaning that whatever affects one species also affects many of the other species in that community. This balance is maintained by a network of interactions between the different species that are part of the ecological community. There are four main types of species interactions that take place in ecological communities: mutualism, commensalism, competition, and predation.

Another term for an ecological community is a biome. It may also be called an ecosystem. All the different species of plants, animals, and insects that live in that community interact with each other to create a natural balance.

9

In mutualism, both species benefit from their interaction. The earlier example of the relationship between bees and plants is an example of mutualism, because both species benefit from pollination. In commensalism, one species benefits from another by gaining food, shelter, or locomotion, while the other species is neither helped nor harmed. Tree frogs using plants and trees to hide from predators is an example of commensalism.

Competition interaction occurs when two species compete, often for a single, limited resource such as food, water, or shelter. In a competition interaction, both species are negatively affected. An example of competition interaction can be seen in the relationship between lions and cheetahs. These two species hunt the same prey, so neither one benefits from the existence of the other.

A remora, or suckerfish, is a type of fish that attaches to and rides larger fish such as sharks and marine mammals such as whales and dolphins. Remoras feed on the bits of food left over when the host animal feeds. This is an example of commensalism.

Predation is the fourth kind of interaction between species. In predation interactions, one species benefits while the other is negatively affected. Predation occurs when one species uses another species as a food source. This can be anything from a bear preying on salmon to a deer eating the leaves off plants and trees. A parasite, such as a flea or tick, feeding on an animal is also a form of predation.

Biologists use food webs to illustrate the interaction between species in an ecological community. A food web is a representation of these interactions based on the flow of energy from producers (plants) to primary consumers (plant eaters such as deer or cattle) to secondary consumers (predators such as wolves or humans).

CHAPTER FOUR

WHEN GONE IS GONE

Although the sudden and unexplained disappearance of species may cause some confusion and alarm among scientists, the disappearance of a species isn't uncommon. In fact, it's a part of the natural biological processes of life on Earth.

When all the members of a species disappear, it's called extinction. When a species becomes extinct, it's gone for good. While this sounds terrible, extinction sometimes helps life on the planet. Extinction may happen because of a small change in the environment. Species affected by this change may evolve and adapt, becoming an entirely new species in order to survive in the changed environment. When this happens, the old species likely goes extinct and is replaced by the new species.

An asteroid hitting Earth would result in episodic extinction.

The thylacine, also known as the Tasmanian tiger or Tasmanian wolf, was a large carnivorous animal native to Australia, Tasmania, and New Guinea. Soon after the last captive thylacine died in 1936, the species went extinct.

The natural extinction of a species isn't a negative event. It's simply a part of the biological evolution of life on our planet. Living things are constantly changing all around us, and some evolve while others go extinct. This slow and imperceptible process is called background extinction. However, a second kind of extinction is known as episodic extinction. Episodic extinction occurs when there are rapid and extreme changes in the environment. These may include climate change or environmental disasters caused by comets or asteroids hitting our planet. Natural extinction may result in one or two species disappearing from an ecological community, but episodic extinction can result in a mass extinction of all or most species in a geographical area.

CHAPTER FIVE
MASS EXTINCTION EVENTS

Scientists estimate that 99.9 percent of animal species that have ever existed on Earth have become extinct. These extinctions were due to catastrophic environmental changes. There have been five of these mass extinctions in Earth's past.

The first of these mass extinctions, the Ordovician-Silurian mass extinction, occurred about 440 million years ago, during the Paleozoic era. This extinction was likely the result of continental drift and the climate changes that followed. First, an ice age killed off all the species that couldn't adapt to the cold. When the ice eventually melted, the rising sea levels killed off more species.

About 400 million to 375 million years ago, the Devonian mass extinction occurred. Nearly 80 percent of all species were wiped out due to rapid cooling of the air and diminished oxygen levels in the oceans, as well as possible volcanic eruptions and meteor strikes. The Permian mass extinction, the possible result of climate change, volcanic activity, and asteroid strikes, occurred about 250 million years ago. This event, known as "The Great Dying," resulted in the extinction of 96 percent of the species on the planet.

The Triassic-Jurassic mass extinction took place about 200 million years ago and resulted in the extinction of almost half of all life on Earth. It's believed that this mass extinction was caused by volcanic activity, climate change, and changing sea levels. The fifth mass extinction, the K-T mass extinction, happened 65 million years ago. This extinction is credited to the drastic climate changes caused by extreme meteor activity. It's the mass extinction responsible for wiping out the dinosaurs.

Dinosaurs were perfectly suited to survive in the environmental conditions of Earth millions of years ago. But when the environment suddenly changed, they were unable to adapt and went extinct.

While the last mass extinction occurred about 65 million years ago, some scientists believe that there may be a sixth mass extinction event happening right now. What makes this possible sixth mass extinction unique is that we may be responsible for it.

The average rate of natural extinction is about one to five species per year. However, due to human activity, extinctions are now occurring at a much faster rate. Biologists warn us that between 150 and 200 plant and animal species are now going extinct every day, and we're the cause. Because of the pollution caused by fossil fuels and the destruction of native habitats, the extinction rate of species is now 1,000 times greater than the rate of natural extinction. Experts warn that this is possibly more catastrophic than the K-T mass extinction event.

First discovered in the 1950s, the vaquita is the most endangered marine mammal on Earth. These small mammals frequently get caught in fishing nets and die. As of 2019, there were fewer than 20 vaquitas in the wild.

Losing almost 200 species of plants and animals every day might sound like an impossible problem to solve. However, there are ways to prevent species from going extinct. Scientists have created systems to categorize species that are at risk. Under the U.S. Endangered Species Act, as a species loses its habitat or many members of its population, it's listed as threatened. If the species' situation doesn't get better, the species is then classified as endangered. Once it's listed as endangered, a species is in serious danger of going extinct. There are currently over 1,800 species of plants and animals listed as endangered around the world.

CHAPTER SIX

WHAT MAKES A SPECIES ENDANGERED?

When scientists talk about endangered species, they're talking about any organism that's threatened by extinction. There are two main ways a species can become endangered. One way is through the loss of their habitat. The other way is through the loss of genetic variation, or difference.

A habitat is the natural environment in which a plant, animal, or another organism lives. Habitats provide food, shelter, protection, and a means to reproduce. Habitats can change naturally over time due to events such as volcanic eruptions, changing ocean currents, or climate change. Sometimes a species may evolve and adapt to these changes. Other times, these changes can be too extreme, causing the species to become endangered and eventually go extinct. An example of natural habitat loss can be seen in the extinction of the dinosaurs. When Earth's hot, dry climate quickly became much cooler 65 million years ago, the dinosaurs were unable to adapt and they became extinct.

Human activity can also lead to the loss of an organism's habitat in a number of ways. In the Amazon rain forest in South America, for example, thousands of acres of trees and other plants have been destroyed to make room for farmers, ranchers, and the logging industry. In parts of North America, expanding human settlement has severely limited the habitats of species such as the mountain lion, coyote, and grizzly bear.

As human society expands, it intrudes on the natural habitats of plants and animals such as the California mountain lion. Destroying a species' habitat makes it difficult for that species to survive in the wild.

Loss of genetic variation is the second factor that can lead to a species becoming endangered. Think about your family and friends. Do they all look alike? Most likely, some of them have brown eyes and some have blue eyes. Some of them have curly hair, some of them have wavy hair, and some of them have straight hair. These differences are caused by the different ways that our genes combine. Genes are the things that carry the information that tells our cells how to function properly.

The diversity, or difference, caused by genetic variation allows a species to adapt to environmental changes. The greater the population of a species, the more genetic variation there is. If there isn't enough variation, a species may lose the ability to evolve and adapt. Variation may occur in one of three ways. First, a mutation may occur. A mutation is a change in the DNA of an organism. Over time, this change may be passed on to other members of the species, allowing the species to evolve. The second source of genetic variation is called gene flow, which is the movement of genes from one population of a species to another. The third source, sexual reproduction, contributes to genetic variation by introducing new combinations of genes into a population. The loss of genetic variation may occur naturally. However, human activity such as overhunting or overfishing has led to a drastic decrease in the populations of certain species, contributing to the rapid loss of genetic variation among these species.

> A famous example of genetic variation is Darwin's finches. Charles Darwin, an English naturalist, found 15 different species of these finches on the Galápagos Islands. The finches evolved from a common ancestor species, in part through genetic variation.

CHARLES DARWIN

CHAPTER SEVEN

HUMAN IMPACT

 While it's easy to see how overhunting or destroying large sections of the South American rain forest can lead to plants and animals becoming endangered, humans are having an impact on the health of species in small and hard-to-see ways as well. In addition to destroying about 1 percent of the world's tropical forests each year, humans have also affected the world's wetlands. Wetlands include lake, pond, river, swamp, and marsh habitats. Many wetland habitats in Europe and North America have been drained over the last two hundred years to make way for human settlements. Changing these habitats has upset the breeding and migration patterns of many species of water birds.

Logging in Malaysia has led to deforestation.

The eastern barred bandicoot is native to Australia. Human activity has led to this species becoming endangered through the destruction of its habitat and the introduction of invasive species such as rabbits and deer.

Another impact humans have on habitats is the introduction of invasive species. An invasive species is a plant or animal that's not native to the habitat. Invasive species can have a very harmful effect on native species. For example, in 1859, several rabbits were introduced to Australia so people could hunt them. In less than 100 years, the rabbit population exploded to 600 million. These rabbits took over the resources of native species such as the bandicoot, which has become endangered.

Human activity also produces pollution. In October 2019, a leak in the Keystone Pipeline in North Dakota released more than 9,000 barrels of oil, poisoning the environment around the pipeline. Burning fossil fuels, such as coal and petroleum, releases poisonous chemicals such as sulfur dioxide and carbon dioxide into the atmosphere.

CHAPTER EIGHT

SAVING ENDANGERED SPECIES

Once a species is extinct, it's gone forever. But the good news is that endangered species can be saved and there's a lot you can do to help. The key factor in saving an endangered species is reducing human impact on the environment.

The first step in trying to save endangered species is education. Educate yourself, your family, and your friends about the endangered species in your area. Maybe you live near a wildlife refuge or national park. Visit these protected lands and learn more about the habitats of native wildlife and plant species. One of the easiest ways to protect endangered species is to protect the places they live. You can also volunteer at a nature center or wildlife refuge and help the people who work there protect the land and the species that call it home.

Next, you can look around your home and see if it's wildlife friendly. Make sure that garbage is kept in cans with locking lids so that animals such as raccoons and bears aren't drawn to it. Feed your pets inside or in fenced-in areas so that wild animals don't get used to eating the food you leave out for your dog or cat. Allow native plants to grow in the area, providing food and shelter for native wildlife.

Make sure birdbaths are cleaned often to prevent the spread of disease, and only fill bird feeders with food eaten by local species. The wrong kind of food may attract invasive bird species and harm the species native to your area.

Volunteering to help clean up local parks and wildlands is a great way to help protect the plant and animal species that live in your area.

25

The one thing we can all do to help the various species we live alongside is to live sustainably. Sustainability is living in a way that makes it possible for human civilization to coexist with the natural world. Sustainability is often called "living green" or "living a green lifestyle." One of the hallmarks of sustainability is reusing and recycling. Recycling paper products means we need to cut down fewer trees. Reusing plastic products reduces the amount of trash that we produce, which means we need to use less land for landfills.

But working to save endangered species can go beyond living sustainably. You can speak out about suffering species by becoming an activist. In fact, many young people are at the forefront of the fight to save endangered and at-risk species.

BLUE-FOOTED BOOBY

Will and Matthew Gladstone of Massachusetts, aged 15 and 12, are two brothers who created The Blue Feet Foundation. This organization is designed to teach people about the blue-footed booby, a declining bird species. The Blue Feet Foundation raises money to protect wildlife by selling bright blue socks.

Carter and Olivia Ries of Georgia were just 8 and 7 years old when they created One More Generation, a group devoted to educating the public about endangered species.

Hannah Testa of Georgia is a teen sustainability advocate. An advocate is someone who argues for a cause or policy. Inspired to fight for sea creatures who were suffering as a result of plastic pollution, Hannah uses her voice to spread awareness about plastic pollution and its impact on wildlife.

> Josiah Utsch, pictured here, was just 11 years old when he started Save the Nautilus. Together with his friend Ridgely Kelly and Dr. Peter Ward, he works to teach the public about this at-risk species.

CHAPTER NINE

THE ENDANGERED SPECIES ACT

In 1973, the United States Congress passed the Endangered Species Act (ESA). The Endangered Species Act gave the federal government the authority to protect endangered species, threatened species, and critical habitats. The Endangered Species Act lists protected plant and animal species in the United States and around the world. Species that are protected by the Endangered Species Act are said to be listed species. There are many other species that are under consideration for possible protection under the Endangered Species Act. These species are called candidate species.

The goal of the Endangered Species Act is to make endangered and threatened species populations healthy so that they can be removed, or delisted. Two government agencies enforce the Endangered Species Act. The U.S. Fish & Wildlife Service oversees the protection of all terrestrial animals and plants, as well as freshwater fish. Fish and wildlife in the oceans are protected by the National Marine Fisheries Service, which is part of the National Oceanic and Atmospheric Administration (NOAA).

Under the Endangered Species Act, it's illegal to harm, hunt, shoot, trap, collect, or capture a listed species. It's also illegal to interfere in the breeding and other natural behavior of a protected species. Listed plant species are protected if they're on federal land. The federal government can also refuse to issue a federal permit to disturb listed species that are growing on private land.

In August 2019, the Trump administration made changes to the Endangered Species Act, weakening protections for endangered species and their habitats. Many critics said this move favored business over the environment and further endangered many species.

CHAPTER TEN
THEY CAN BE SAVED!

There are solutions to the endangered species problem. Scientists and organizations such as the World Wildlife Federation and the National Audubon Society are working hard to help save endangered and threatened species around the world. One way to protect endangered and threatened species is through conservation. Conservation is the practice of protecting endangered species and their habitats in order to prevent them from going extinct. Conservationists not only work to stop people from harming or killing endangered species, they also work to protect their habitats from being destroyed.

There are a number of species that no longer exist in the wild but are not yet extinct. The sole surviving members of these species live in zoos and nature reserves around the world. Zoos give them a safe place to live and breed, increasing the population of the species. Once the young animals are old enough, they can be reintroduced to their natural habitat, increasing the species population in the wild.

Two of the most noteworthy conservation successes are the giant panda and the California condor. The combined conservation efforts of Chinese and American scientists have helped bring the giant panda back from the edge of extinction. In fact, the giant panda's status has been changed from "endangered" to "vulnerable." At one time, there were only 27 California condors left on the planet. Thanks to the efforts of the San Diego Wild Animal Park and the Los Angeles Zoo, there are now hundreds of condors in the skies over California.

GLOSSARY

activist (AK-tih-vist) Someone who acts strongly in support of or against an issue.

Alzheimer's disease (AHLZ-hy-merz dih-ZEEZ) A disease of the brain that causes people to slowly lose their memory and mental abilities as they grow old.

categorize (KAA-tuh-guh-ryz) To group or class things.

catastrophic (kaa-tuh-STRAH-fik) Related to or like a terrible disaster or something that happens suddenly and causes much suffering and loss for many people.

continental drift (kahn-tuh-NEN-tuhl DRIFT) The very slow movement of the continents on the surface of Earth.

describe (dih-SKRYB) To explain.

DNA (DEE EN AY) A matter that carries genetic information in a plant or animal's cells.

diminished (duh-MIH-nisht) Became or caused something to become less in size.

drastic (DRA-stik) Extreme in effect or action.

fossil fuel (FAH-suhl FYOOL) A fuel—such as coal, oil, or natural gas—that is formed in the earth from dead plants or animals.

imperceptible (im-puhr-SEP-tuh-buhl) Impossible or very hard to see or notice.

indigenous (in-DIH-juh-nuhs) Having started in and coming naturally from a certain area.

migration (my-GRAY-shuhn) The act of an animal or people moving from one location to another.

parasite (PEHR-uh-syt) A living thing that lives on or in another living thing and often harms it.

pollinate (PAH-luh-nayt) To take pollen from one flower, plant, or tree to another.

terrestrial (tuh-REH-stree-uhl) Relating to or occurring on Earth.

INDEX

B
background extinction, 13
Blue Feet Foundation, the, 27

C
climate change, 13, 14, 18
commensalism, 8, 10, 11
competition, 8, 10
conservation, 20

D
Darwin, Charles, 20, 21

E
Endangered Species Act, U.S., 17, 28, 29
episodic extinction, 12, 13
extinction, 12, 13, 14, 16, 18, 30

F
food web, 11

G
genetic variation, 18, 20
Gladstone, Will and Matthew, 27

I
invasive species, 23, 24

K
Kelly, Ridgely, 27

M
mass extinctions, 14, 16
mutualism, 8, 10

O
One More Generation, 27

P
predation, 8, 11

R
Ries, Carter and Olivia, 27

S
Save the Nautilus, 27

T
Testa, Hannah, 27
Trump administration, 29

U
Utsch, Josiah, 27

PRIMARY SOURCE LIST

Page 13
Tasmanian "Zebra Wolf" Thylacinus in Washington, D.C. National Zoo. Photograph. Circa 1904.

Page 21
Darwin's finches or Galápagos finches. Illustration. Charles Darwin. 1845. *Journal of researches into the natural history and geology of the countries visited during the voyage of H.M.S. Beagle round the world, under the Command of Capt. Fitz Roy, R.N. 2d edition.*

Page 26
Josiah Utsch, 12 yrs. and Ridgely Kelly, 11 yrs. Photograph. Cape Elizabeth, Maine, December 20, 2012. Accessed through Getty Images.

WEBSITES

Due to the changing nature of Internet links, Rosen Publishing has developed an online list of websites related to the subject of this book. This site is updated regularly. Please use this link to access the list: www.powerkidslinks.com/SOGI/endangeredspecies